Pompeii

A HISTORY OF THE CITY AND THE
ERUPTION OF MOUNT VESUVIUS

Fergus Mason

HistoryCaps
ANAHEIM, CALIFORNIA

Copyright © 2019 by Golgotha Press, Inc.

All rights reserved. No part of this publication may be reproduced, distributed or transmitted in any form or by any means, including photocopying, recording, or other electronic or mechanical methods, without the prior written permission of the publisher, except in the case of brief quotations embodied in critical reviews and certain other noncommercial uses permitted by copyright law.

Contents

About HistoryCaps 1

Introduction 3

Part One: Early History 9

    Rome 12

    The Sea People 19

    Carthage Must Be Destroyed 24

    The Social War 31

    Roma Victor 41

Part Two: First Century Pompeii 47

    The Wealth of Pompeii 49

    A Night on the Town 57

    The Imperial Machine 63

    The Earthquake 67

Part Three: Death in August 81

    The Exploding Mountain 83

    The Fleet 87

    Incineration 92

    The Admiral 97

Discovery and Conservation 104

Conclusion 110

ABOUT HISTORYCAPS

HistoryCaps is an imprint of BookCaps™ Study Guides. With each book, a brief period of history is recapped.

We publish a wide array of topics (from baseball and music to science and philosophy), so check our growing catalogue regularly (www.bookcaps.com) **to see our newest books.**

Introduction

Take a boat out to the middle of the Bay of Naples and you'll have one of the most spectacular panoramas in Italy. Set on the Mediterranean coast of the country's Campania region, the ten mile wide bay is an amazing sight. The blue waters are bustling with activity; Naples is one of Europe's busiest ports and handles more passengers than any other worldwide except Hong Kong. Many of Italy's cruise lines have their bases here, and their luxurious ships are constantly shuttling in and out. Smaller, sleeker gray hulls sometimes slip among them, because the US Navy's Sixth Fleet has its headquarters here too. Ferries, tourist boats, fishermen and small transports add to the endless sea traffic.

To the north and south of the Bay small peninsulas run out into the Mediterranean, pointing at the islands that lie beyond them. The islands themselves are lively. In the north, tiny Procida with its fishing ports and growing tourism industry separates the spa resort of Ischia from the mainland; to the south lies beautiful Capri, one of the region's most popular summer destinations.

Inland of the Bay the land is low and rolling for about ten miles, before rising to the sharp ridges of the Campanian massifs. These thickly wooded slopes are a remnant of the last Ice Age, but now they're scattered with small villages and summer homes where Italians go to relax. Tourists enjoy them too, and their money is a welcome boost to the local economy. The centerpiece of the landscape has to be Mount Vesuvius, though. Its summit reaches up over 4,200 feet from the coastal plain, not high by the standard of the Rockies but more impressive when you remember that its lower slopes end on the shoreline. It's a saddle-backed mountain with two summits – the high peak of Vesuvius itself and, to the north, the ridge of Mount Somma reaching up to 3,770 feet. From anywhere in the Bay, and even far inland, it dominates the skyline. Around its foot an urban sprawl - the Naples metro area - is home to over four million people.

The towns around the Bay cover the whole spectrum of settlements from fishing villages to the city and port of Naples itself. From a distance, any one of them is a picturesque Mediterranean scene, with lots of pastel-pained buildings and red tile roofs. Get closer and a lot of the houses and businesses start to look a bit run down and neglected. Some of those roofs aren't tile at all; they're corrugated iron and they're red with rust. Strangely the tatty appearance doesn't take away from the appeal, though. Everything is infused with a cheerful, exuberant chaos that's typically Italian. The concrete might be crumbling, and half the cars have dents at the corners, but it's a place people are happy to live in.

Of course, just because the bay is humming with life doesn't mean there aren't dangers lurking here. It's one of the most corrupt cities in Europe, and in southern Italy corruption doesn't stop at passing brown envelopes. The feared Camorra, one of the oldest and largest of Italy's organized crime syndicates, dominates the local economy. Crossing them can be lethal, and their "waste disposal" rackets sometimes leave parts of Naples buried in garbage for weeks at a time.[1]

More bizarrely, it's believed that 20 nuclear warheads lie buried in the ooze at the bottom of the Bay. They were planted by a Soviet submarine in January 1970; the plan was that, in the event of war, they could be remotely detonated to destroy the Seventh Fleet HQ and any warships anchored in the bay.[2]

When the Neapolitans found out about the warheads, there was a distinct lack of the fuss you would normally expect. It would be natural to see some panic among people who've just been told there are 20 rusting Soviet nukes on their doorstep, but by and large the people of Naples just shrugged and went back to their espresso. Amazing as it might seem, if you live on the Bay of Naples a few stray weapons of mass destruction buried somewhere out on the seabed aren't what you need to worry about. The real danger is standing in plain view.

From the Bay, the summit of Vesuvius looks like a rounded cone, but it isn't. It's a partially collapsed hollow dome, its broken top revealing a thick-walled rock chamber floored with steaming sand. Vesuvius is an active volcano, a geothermal bomb of indescribable power. In the last 2,000 years, it's erupted 56 times, three of them in the 20th century. The last 55 eruptions have been relatively small but even then some killed hundreds of people, and one in March 1944 destroyed most of a USAAF bombardment group's aircraft. That was the last eruption, and for nearly 70 years Vesuvius has been quiet. That's a bad sign. Deep below the mountain gases are boiling off inside a reservoir of molten rock, steadily raising the pressure on the upper slopes towards an unguessable breaking point. One day – nobody knows when – Vesuvius will unleash a massive pulse of fire and rock, and scythe the whole coast of the bay clear of life.

    It's done it before.

[1]
Part One: Early History

In the centuries before the rise of Rome Italy was a crazy patchwork of kingdoms. The population was a mixture of Italic peoples and others from outside Europe, such as the Etruscans. The kingdoms were often hostile to each other but at the same time they traded widely, with each other and with tribes outside Italy. The map was never static for long, as peoples emerged, split and faded away. Early in the first millennium BC an Italic group known as the Osci arose in the southern half of the Italian mainland. Their origins aren't well known, but relics of their settlements have been found throughout what's now the Campania region, and together with later records it's possible to piece together a lot of details. Some time in the 7th or 8th century BC a group of Osci founded a settlement on the west coast, in a magnificent bay where Greek and Phoenician ships sometimes sheltered in bad weather. It was an ideal place for a port. It lay at a crossroads between Nola, Stabiae and the Greek enclave of Cumae, so the inhabitants could trade for goods with foreign sailors then market them to the surrounding towns.

The port seems to have prospered over the next couple of centuries, but in the hostile political environment of Italy that prosperity attracted attention from outsiders. The Etruscans conquered it in the 6th century,3 and ruled it long enough to create their own cemetery and leave inscriptions. Within a few decades, it was captured again, this time by the Greeks from Cumae, and in the 5th century BC it fell to the Samnites. The Samnites were closely related to the Osci and spoke the same language; it's likely that the town, by then known as Pompeia, returned to something close to its original culture despite being enlarged by its new rulers.

Rome

Things were changing in Italy though. One of the region's many city-states, Rome had been founded near a ford on the Tiber River in the mid-8th century BC and had been growing slowly ever since. The Etruscans had dominated it from the late 7th century, but they never imposed direct rule; Rome had its own king and was left undisturbed to develop a distinctive political system. The Senate was founded almost as soon as Rome became large enough to need a government, and consisted of a panel of the town's most respected citizens. It had remarkably little real power as that was mostly concentrated in the hands of the king, but it at least looked like a real system of government.

The influence of the Etruscans slowly faded while Rome continued to grow. The seventh king of the city, Lucius Tarquinius Superbus, launched a series of successful wars against neighboring states. He also ordered several large public works projects, including the city's sewer system and an expansion of what would eventually become the largest sports stadium in human history - the Circus Maximus. Tarquinius had few friends though, and he quickly created a large number of enemies for himself. He'd seized the throne of Rome from his predecessor (and father in law) Servius Tullius, then had him murdered. He had favored the rich over the ordinary people of Rome, and used violence and intimidation to stay in power. He also ignored the city's traditions, disregarded the Senate and broke the law with impunity. Finally, his son Sextus raped Lucretia Collatina, a woman from a powerful family, then tried to threaten her into keeping silent. Instead, she called her family together, told them what had happened then committed suicide with a dagger.4

Lucretia's relatives were outraged at the crime but knew the king would protect his son. The king was out of the city, though, and the government was vulnerable. Lucretia's husband Lucius Tarquinius Collatinus and his cousin Lucius Junius Brutus – both of whom were also relatives of the king – led a group of four who barred the city gates, expelled Sextus and formed a republican government based on the Senate. The new government was led by two consuls, each elected by the Senate and each able to cancel any instruction given by the other. It was an early form of checks and balances, designed to prevent any one man being able to hold absolute power, and it would be in place for over 450 years. The Roman Republic had been born.

The Republic was more stable than its neighbors, and more able to follow a long-term goal. It was also able to defend itself effectively as a trained citizen army began to emerge. Gradually Rome began to have more of an influence on the constant wars that erupted throughout central Italy, even when its opponents looked far stronger on paper. In 343 BC the Samnites, by this time a large and powerful confederation who controlled much of central Italy, launched an attack on several tribes in northern Campania. They quickly overran several small kingdoms of the Campani and finally besieged the wealthy city state of Capua. The Campani, out of options, turned to Rome for help. The Romans had a treaty with the Samnites, however, which set the Liris River as a boundary between their areas of influence, and the Campani were on the other side of the river. The Senate were tempted by the thought of an alliance with the wealthy Campani but to honor the treaty they had to refuse the request. The refusal ended the Campani's last hope of victory against the Samnites, so now they played their final card – they surrendered unconditionally. To Rome.

That changed everything. By surrendering the Campani had become subjects of Rome and their land was the Republic's property. Roman honor would not let the Samnites continue their attack whatever the treaty said, so envoys were sent to the Samnites to explain the new situation and request that, in view of the friendship between the two powers, the Campani lands be left alone. The message ended with four letters that probably meant nothing to the invading tribesmen, but would come to be the symbol of ultimate power through most of the ancient world. Senatus Populusque Romanus, the Senate and People of Rome, were making the request. As the Samnites would soon learn, any demand marked SPQR could not be lightly ignored.

Rome wasn't what it would become though. It was a small state centered around a single city, and while its neighbors were learning that the well drilled Roman troops could be a useful ally nobody seriously feared the tiny Republic. In fact the Falerii had been maintaining an uneasy truce with Rome for 40 years, refusing to sign a permanent peace agreement, and the Latins were actually planning to attack the Republic. Now the Samnites thought they could ignore the bluster of the Senate and People; their magistrates ordered a renewed attack on

Capua, and made sure the Roman envoys heard them do it.

Rome might have been small but it was determined, and the Samnite refusal spurred the Republic into action. Both consuls took command of armies and marched against the Samnites. One headed into Campania to break the siege of Capua; the other invaded Samnium itself. The Samnites were happy to accept the challenge but it didn't turn out the way they expected. The Romans defeated the attacker's armies in a series of three battles and the Samnites backed off. The Campani became a Roman ally and the Falerii, startled at Rome's military performance, hastily agreed to a formal peace treaty. The Latins decided to find an easier target and attacked the Paeligni instead. Suddenly Rome was a notable force in central Italy and other states began to tread more warily around them. They still weren't a dominant power, or even first among equals, but their neighbors began to take them seriously. Rome's influence had expanded too and many towns which had been loyal to the Samnites were adjusting to the new reality. One of them was Pompeia. The city – and the rest of Campania – remained largely independent, but their paths were increasingly tied to Rome. That had benefits for a cash-rich but militarily weak state, but it brought dangers too. When a second war erupted between Rome and Samnium in 310 BC a party of Roman marines landed near Pompeia and looted the town.5

The Sea People

Rome was gaining power on the Italian peninsula but the undisputed superpower of the western Mediterranean was Carthage. Founded in 814 BC as a dependency of the state of Tyre, it became independent around 650 BC. Like Tyre it was a Phoenician city and that shaped its future rise. The Phoenicians were descended from the Canaanites who had inhabited the southeastern shores of the Mediterranean since around 8,000 BC. In the late second millennium BC the Canaanites started to diverge into distinct peoples. The Israelites emerged around 1,300 BC and dominated the area that is now Israel and the West Bank. The kingdoms of Aram-Damascus, Ammon, Moab and Edom occupied the higher ground between the Jordan River and the borders of the Assyrian empire. These were all land powers and their constant squabbles prevented any of them from becoming extremely significant. The Phoenicians controlled only a tiny strip of land on what's now the Lebanese coast and to make up for their shortage of territory they turned to the sea. Once on a ship they weren't hemmed in by jealous kings and aggressive priests, and they ranged the length of the Mediterranean. Carthage, although founded by Tyre, lay 1,400 miles to the west of it.

Carthage broke with Tyre in the 7th century BC and set about some expansion of its own. Tingis (now Tangier), established in the late 6th century BC, marked the western limit of Phoenician civilization – it's on the Straits of Gibraltar, 900 miles from Carthage. After founding Tingis the Carthaginians looked out into the cold Atlantic, decided it looked too threatening for their elegant, fast rowing galleys and turned north instead. Crossing the Straits was easy for them and their expansion picked up speed once more. This time they moved back east, trading with the peoples they met and establishing their own cities as they went. By the 4th century BC their possessions included the whole southern coast of Spain, all the greatest western Mediterranean islands – the Balearic Islands, Corsica and Sardinia – and most of Sicily. It was Sicily that finally led to their downfall.

After the war with the Samnites Carthage had sent Rome a gold crown to congratulate them on their victory. Over the 80 years since then Carthage had continued to expand through the western Mediterranean and Rome had increased its territory in Italy, but there had been few clashes between them. Rome's strength – including its developing army – was all on land, while Carthage was almost entirely a sea power that relied on mercenaries when an army was required. There was little potential for conflict between them and relations were generally good. In 281 BC, however, a minor war flared up between Rome and the state of Tarentum. It quickly expanded as both sides' allies got dragged in. Tarentum turned to the powerful Greek colonies in southern Italy, while Rome sided with Carthage.

The Roman army had been reforming and no longer used the Greek-style Hoplite formations the Roman Kingdom and early Republic had copied from the Etruscans. The troops were now formed into units of about 5,000 men, the legions. Their organization and tactics were still crude compared with what they would become but already their army was proving itself against the other Italian states. They were still a local power though, and hadn't got involved in the wider affairs of the dominant Greek and Phoenician states. Now Tarentum was bringing in mainland Greek allies and they included the ambitious king Pyrrhus of Epirus. Pyrrhus saw a chance to take land in Italy, perhaps including Rome itself. It didn't work out that way though. Pyrrhus won victories but they were costly – the term "Pyrrhic victory" comes from the words of Pyrrhus when a follower congratulated him on defeating a Roman army: "If we are victorious in one more battle with the Romans, we shall be utterly ruined."[6] The problem was that the Greeks, more used to fighting major wars, were winning the battles, but their army was only 25,000 men. The Romans had ten legions, more than 50,000 men. They could be defeated time after time, but they simply reorganized themselves and attacked relentlessly. In 278 BC Pyrrhus took his army to Sicily to help the Greeks there fight the Carthaginians, and the legions immediately destroyed his allies in southern Italy. Finally Pyrrhus returned to the mainland in 275 BC for one final battle against Rome. This time the Romans, more experienced after five years of war, managed to fight the Greeks to a standstill. Pyrrhus saw the writing on the wall and left Italy forever, leaving the Romans to overrun the last Greek colonies in the south.

Carthage Must Be Destroyed

Rome and Carthage had been allies through the Pyrrhic War, but now they were facing each other across the Straits of Messina which separated the legions in Rome's new territory from the Carthaginian colonies in Sicily. It didn't take long for a new conflict to erupt. In about 290 BC the Mamertines, a group of Italian mercenaries, had captured the town of Messana (now called Messina) and turned it into a base for raiding and piracy. In 270 BC a Greek army from Syracuse landed and attacked the Mamertines, who had been plundering the remaining Greek towns in Sicily. First the Mamertines asked for Carthaginian protection, then they changed their minds and appealed to Rome. The army of Syracuse immediately formed an alliance with the Carthaginians, and this forced Rome to act. Nervous at their new role as a major player in Mediterranean geopolitics, and unwilling to have a potentially hostile power on Sicily, they sent a garrison to occupy Messana. Tension quickly rose and in 264 BC war broke out between Carthage and Rome. The first significant battle didn't take place until 262 BC, when a Roman army defeated a Carthaginian one in western Sicily. Carthage decided at this point to avoid confronting the power of the legions and fight the war at sea. The small Roman navy was defeated in 260 BC, but instead of accepting Carthaginian naval superiority Rome instead began building a vast new fleet based on the design of a captured Carthaginian galley. The speed of this expansion was incredible – 120 warships were built in only 60 days. By using legionaries as marines they quickly gained the upper hand at sea as well. In 241 BC, after a string of defeats, Carthage signed a peace treaty that saw them handing

Sicily to Rome along with a whopping cash payment. Over the next few years, as a weakened Carthage tried to suppress rebellions among its unpaid mercenary army, Rome occupied Corsica and Sardinia as well. The Carthaginians were furious and wanted revenge. In 219 BC they tried to get it.

Hannibal, one of the leading Carthaginian military commanders, began by attacking Rome's allies in Spain then led an invasion force through the Alps into Italy. He repeatedly defeated and destroyed Roman armies, including at the battle of Cannae in 216 BC where over 50,000 Roman soldiers died. His strategy failed though; most of Rome's allies didn't change sides, as he'd hoped (although Capua did) and he failed to take the city of Rome itself. Most of his famous elephants had died on the march through the Alps and he had no siege equipment to break through the city walls. A long siege might have starved the defenders out but that wasn't possible. No matter how many Roman armies were destroyed new legions were always raised and thrown into the fight. Hannibal spent 15 years trying to achieve a decisive victory, while the Carthaginian armies in Sicily and Spain were defeated. In the end Rome's resilience was too much for him and he was bottled up in the south of Italy, reduced to pillaging farms. Finally the Roman fleet landed yet another army in North Africa, threatening Hannibal's supply base. He rushed back with his own army to defend it and found himself up against an improvised Roman force of 35,000 men, built around a core of veteran legionaries and led by Publius Cornelius Scipio. Hannibal had 40,000 men and believed he could win, but he was wary of Scipio's military reputation. Nevertheless on October 19, 202 BC he formed his army up on the plain of Zama and attacked. It was a disaster; Scipio's men smashed the three lines of Hannibal's army one by one, killing half his troops and capturing most of the rest. The Romans lost only 2,500 men. This shattering victory ended the second of what the Romans called the Punic Wars.[1] As well as making another large payment to Rome the Carthaginians were forced to give up al-

---

[1] The Latin word for *Phoenicians* is *Punici*.

most all their territory, leaving them only a small area around Carthage itself. With their enemy defeated the Romans turned their attention back to Italy. It was time to reward their allies who had stayed loyal, including Pompeia, and punish the few who had joined the invaders.

Capua had switched sides after the Roman defeat at Cannae, and Hannibal himself had used the wealthy city as a base. In wealth and influence it was one of the leading cities in Italy, second only to Rome itself, and the Republic didn't take its defection well. Several attempts were made to capture it, which Hannibal's army always beat off bloodily. In 211 BC Hannibal outsmarted himself however, and the city fell to the legions. It was immediately made Roman state property, the city senate dissolved and the inhabitants brought under Roman rule. Parts of the city were sold while the war was still going on. After Hannibal's defeat at Zama more was sold and the rest rented out to Roman businessmen. Capua remained wealthy but its political influence was broken. The towns of Campania now increasingly looked to

Rome for leadership. Pompeia, which had remained loyal throughout the war, became more prosperous after it. As a major port it attracted a lot of shipping, and now it had a privileged position as a trading center for the leading power in the Mediterranean.

Rome's power was set to grow even more. Carthage had been crushed in the Second Punic War, but slowly rebuilt its military to defend against raiding Nubian tribes. This wasn't acceptable to Rome, which now viewed Carthage as a client state that should do what it was told. Leading politician Marcus Porcius Cato (now known as Cato the Elder) took to ending every speech, no matter what it was about, with the words Carthago delenda est – "Carthage must be destroyed."[7] The Senate agreed and in 149 BC Rome began provoking Carthage with a series of ever more outrageous demands. Carthage was desperate to avoid another war but when the Romans told them their city was too close to the sea, and demanded that they demolish it and rebuild it further inland, they'd had enough. They refused. That, of course, was just what Rome had been waiting for. An army led by Scipio Aemilianus besieged Carthage. In the spring of 146 BC they broke through the walls, annihilated the remains of the defending army and burned the city to the ground. Of half a million people who had lived in Carthage only about 50,000 survived; they were sold as slaves and what had been Carthage became the Roman province of Africa. That marked the start of Rome's expansion on the southern shore of the Mediterranean, and within a few years huge quantities of grain and other goods were being carried north to Italy. Much of it came ashore at Pompeia. The town, still independent of Rome but increasingly tied to the Republic's economy, began to grow rich.

The Social War

Pompeia was now an ally of Rome, but the Republic's allies could be divided into two classes. The Latins and a few others had willingly associated themselves with Rome and, while increasingly reduced to client states, were respected and given most of the benefits of Roman citizenship. Former enemies who had fallen under Roman influence, the Socii, didn't get as much respect. They paid taxes to Rome and their people could be conscripted into Roman armies, but they didn't have any influence over the Republic's policies. In 91 BC many of the Socii in Campania rebelled against their Roman overlords. Pompeia joined them. It was a disastrous error.

The Socii wanted independence from Rome and planned to create a new confederation of states called Italia. Now they created a currency for this new union, and chose a capital city – Corfinium, which they renamed Italica. It might seem as if they had their priorities wrong, because creating symbols of statehood was less influential than eliminating the threat of the Roman army, but in fact the Socii were confident. Twelve states were involved in the rebellion and between them they had an army of about 100,000 men. Most of the soldiers were veterans of the legions themselves, so it was a powerful and battle-hardened force. The rebellious cities believed it was powerful enough to take on the Romans and win. This confidence seemed justified; Rome's own army was larger, but not by much, and it couldn't all be used against the rebels. Troops were needed to defend cities, secure supply routes and help keep other Italian states loyal or neutral.

In the first year of the war the Socii won a series of victories against small Roman forces, but couldn't force a decisive battle. Rome's two consuls each took command of a strategic direction – one in the north and the other in the south – and concentrated on defending towns and channeling the Italian armies away from Rome. In 90 BC the consul commanding the northern army was defeated and killed by an Italian army, but his military adviser managed to lead most of the army out of the trap and took sole command. His name was Gaius Marius, and thanks to him the rebellion of the Socii had been doomed before it even began.

The Roman army that most of the Italian soldiers had served in had been well trained and effective, but it wasn't a full time force. Legions were recruited when needed, and often fought for a season then disbanded so the men – most of them conscripts – could go home to harvest their crops. The army was only open to those who owned property and there was a clear distinction between Roman legions and those recruited from allied states. At the same time weapons, equipment, training and discipline varied between legions. In fact legionaries had to supply their own weapons, which was one reason for recruiting only from property owners – it shifted the cost of armaments from the Republic to the soldiers themselves. The problem was that the variety of weapons in use made it difficult to develop sophisticated tactics.

Gaius Marius became a consul in 107 BC and immediately became responsible for fighting a war in north Africa. He faced a monumental obstacle, however - he didn't have an army and couldn't find enough eligible recruits to form one. Almost everyone who was qualified to become a legionary was already off fighting elsewhere. That obviously wasn't acceptable and Marius, with the permission of the Senate, began an ambitious series of reforms that would transform the Roman army and, finally, most of the ancient world.

The first stage was to abolish the property requirement; this opened up a huge new pool of manpower from the lower classes of Roman society. Next all Italian soldiers in the army were automatically granted Roman citizenship. That change removed the distinction between Roman and allied legions and made the army more unified. Then Marius went further. He dramatically reduced conscription and created a standing army. Now legions wouldn't be raised when needed and disbanded again when the war ended; they would be permanent military formations manned by professional soldiers. Terms of service were introduced that wouldn't look unfamiliar to a modern western soldier. Now a man enlisted in the army for a fixed term, initially 16 years (it later rose to 20.) If he completed that term he received retirement benefits, a pension and a plot of land in territory his legion had conquered. As well as ensuring commitment and loyalty this also gave the legionaries a large incentive to seize land. While a soldier looked forward to his retirement he received a monthly wage, and could earn promotion to higher ranks that brought better pay and higher status. In return he was expected to pass proficiency tests, including tests of physical fitness, and take part in regular training exercises. Finally the practice of men supplying their own equipment was ended; now every legionary was issued standardized armor and weapons, and new tactics were developed to take advantage of this. In fact what Marius created was a true professional army of a sort that wouldn't be seen again until the Napoleonic wars.

Many of the Socii's soldiers had served in these new model legions, but the full impact of the reforms took years to emerge. More importantly, one of the secrets of its effectiveness was the stability troops got from belonging to a permanent unit. The hastily formed armies of the Socii might be made up of well trained soldiers but they didn't have the team spirit and ferocious loyalty found in a legion. Another factor, which now gave Rome a massive boost, was that recruits integrated into an existing unit were far more effective than if they were formed into a new one. Meanwhile Rome was recruiting every volunteer it could find, and there were many. Attracted by the promise of wages, loot and citizenship men flocked in from all over Italy to put on the red tunic of a Roman soldier. The ranks of the legions swelled to full strength and beyond. Throughout 90 BC the Romans held firm and mostly succeeded in holding off the Italian attacks. In 89 BC they decided it was time to turn the tables. They formed up their new professional army and unleashed it on the rebel states. The Socii were quickly forced back by the onslaught, and city after city fell or was besieged. Pompeia was surrounded by an army led by one of Rome's leading generals, Sulla. Other states among the Socii sent troops to help Pompeia, because the port was a vital link to potential allies outside Italy, but the siege could not be broken.

Lucius Cornelius Sulla is one of ancient Rome's most interesting figures and played a crucial role in politics during the late Republic. After the Social War he was elected as a consul, but in late 88 BC a constitutional crisis provoked him into marching on Rome with six legions – the first time this had ever happened – and forcing reform of the Senate. Then he went off to war again. In 83 BC he marched on Rome a second time; now he used his loyal legions to overthrow the consuls and install himself as Dictator, a position that hadn't been used in over a century. For two years he held absolute power, but unlike most of the dictators who have followed him he concentrated on constitutional reform rather than enriching himself. Then in 81 BC he stunned the civilized world – he resigned and handed power back to the Senate. Next year he stood for election as consul again, and won. One more term was enough for him though; he was an old man now, and he retired to his villa in Puteoli (modern Puzzuoli) where he died in 78 BC.

In 89 BC Sulla was at the height of his military powers. Described by a rival as having the cunning of a fox and the heart of a lion, he was one of Rome's most brilliant generals. Now Pompeia was hammered by the full power of six legions. It was a tough objective; the entire town was surrounded by a 30 foot buttressed stone wall, and shortly before the war began it had been resurfaced to protect against the impacts of Roman artillery. Defensive towers had been built around it and the gates were reinforced. Much of the stone used in the work was volcanic tuff and lava, thrown out in the distant past by Vesuvius. Now more stones were being thrown by Sulla's siege artillery. Onagers – long range catapults that took their name from a species of wild donkey – hurled giant boulders into the town.8 Ballistae and smaller catapults pelted the wall with missiles and hammered the gates. Auxiliary slingers, local militias who accompanied the legions and provided specialist skills but weren't regular soldiers, moved in close to the city and rained lead slugs on the defenders.9 Out of range of slingers and archers on the walls Sulla's legionaries made storming ladders and prepared for an assault.

Not all the records of the Social War survive, so it's impossible to be sure what happened next, but we can guess. On his approach to Pompeia Sulla had attacked several other Campanian towns, including Nola and Stabiae. The destruction was atrocious; according to Pliny the Elder, Stabiae could no longer be recognized as a town after Sulla was finished with it.10 It's likely the citizens of Pompeia heard of the carnage from refugees and decided to spare their own city by surrendering. In any case the siege ended not long after it began and Pompeia came back under Roman authority. For nine years it had much the same status as it had before the war, then in 80 BC Sulla used his consular powers to seize it as a

Roman colony and give much of the land round the town to his veterans. Citizens who had been leaders of the rebellion were expelled from the town, and their property confiscated. For Pompeii everything – even the name – had changed. Now it was part of the Roman Republic and its inhabitants were Roman citizens. There was tension between the original inhabitants and Sulla's veterans who had settled the area,11 but gradually this died down and the town began to thrive again. The volume of trade to and from Rome was greater than ever, both through the port and along the Appian Way.

Roma Victor

The trappings of Roman culture began to appear in Pompeii. Between 80 and 70 BC an amphitheater was built in the eastern corner of the city. It is the earliest known stone amphitheater, and it's still a striking sight. An elliptical oval arena is ringed by rising banks of stone seating that from a distance has an uncanny resemblance to a modern sports stadium. Low walls separated the best seats in the first five rows from the cheaper ones behind and above them. The outer wall of the structure supported arms for a canvas awning that could be extended over the seating to protect against rain or strong sunshine, and private boxes along the wall were reserved for upper class women who wanted to watch the games discreetly.12 The Pompeii amphitheater is the earliest known example of politicians promoting sports to buy popularity; Gaius Quintius Valgus and March Porcius paid for its construction, and had a prominent inscription carved on the gate to make sure nobody forgot. It is also the first known venue of a riot between rival fans, pre-dating English soccer hooligans by over 2,000 years. During a gladiator fight on July 10, 59 AD Pompeians started trading insults with fans from Nuceria, and a fight broke out. It quickly escalated until swords were used and several of the visiting fans died.13 When news of the riot reached Rome the emperor Nero ordered an investigation, and later banned gladiatorial fights in Pompeii for a period of ten years. The amphitheater wasn't closed entirely though. Nero's wife Poppaea, whose family came from Pompeii, persuaded him to leave it open for beast hunts and athletic competitions. Pompeii was a good town to be an athlete in; as well as the amphitheater itself there was a large training ground nearby which included a swimming pool.

Larger and greater construction projects were begun. One of the wonders of Rome, a technological advance that stunned visitors from other states, was its system of aqueducts. Most city dwellers throughout the ancient world drew their water from wells, or from rivers that flowed through the town. These rivers were usually severely polluted with sewage and disease regularly decimated urban populations. In fact dirty water was one of the factors that prevented cities growing too large. Even wells weren't safe; filth and sewage littered the streets and soaked down into the water table, polluting the groundwater. Cholera and dysentery were rife. Rome reached the limits of its local water supplies in the mid-4th century BC, and following the rules of epidemiology should have stopped growing. The Romans simply refused to accept that though. They needed more water and they set out to get it. Central Italy had plenty of water, but it wasn't in Rome. That, the senators decided, was no problem; they'd simply bring it to Rome. The first aqueduct, the Aqua Appia, was built in 312 BC. It ran in from the hills ten miles east of the city and sloped gently down at a constant, almost imperceptible angle for its whole length; the total drop was only 33 feet. Every day it delivered 75,000 tons of fresh spring water to the city. The new channel enabled the city to grow larger, and from then on every time population outran the water supply the engineers just built another aqueduct. The Aqua Aniene opened in 270 BC, bringing in twice as much as the Aqua Appia. The Aqua Marcia followed in 140 BC and the Aqua Tepula in 127. By the time Pompeii fell to Sulla's legions Rome's water system was bringing in over 500,000 tons of water daily and distributing it to every neighborhood of the city, supplying cisterns and tanks in every public square and every marketplace. Rome's population had grown to a million people. The average modern American uses 98 gallons of water per day,[14] far more than any other nation.[15] Each citizen of late Republican Rome could use 132 gallons. Even by today's standards the scale and efficiency of the Roman water supply is formidable. More than 2,000 years ago it was almost beyond the human imagination.

The Republic's endless political wrangling finally brought it down in the late 1st century BC. In 27 the Senate gave full power to Augustus and the Republic became the Empire. Pompeii was part of the Empire and could expect to share its benefits. It did. One of the first Imperial public works projects was a prodigious aqueduct to supply the towns around the Bay of Naples, and in 20 BC a spur was run to Pompeii. 4,000 tons of water a day roared into the Castellum aquae, the central water tank that supplied the town's cisterns, baths and fountains.16 The flow wasn't as extravagant as the supply to Rome but with a population of around 15,000 it still gave each resident almost 70 gallons daily. That's higher than the domestic consumption in every country today apart from the USA and Canada.

Pompeii prospered and grew. The people became thoroughly Romanized. Memories of the Social War gradually faded as that generation passed, and the bustling port city was absorbed into the growing Empire. The military had returned to Pompeii in 27 BC, but not as occupiers this time. Now the Bay of Naples was home to the classis Misensis, the Roman equivalent of the US Navy's Sixth Fleet. This, the most powerful assembly of warships in the entire Mediterranean, operated from a massive new naval base at Misenum on the northern peninsula of the Bay. Sailors and marines spent their wages in the wine shops, restaurants and brothels of the coastal towns. Pompeii collected quite a lot of their money. Businesses thrived, entrepreneurs opened new bars to fleece the navy and merchants built warehouses around the port to bring in the wealth of the Mediterranean. The future of Pompeii looked bright. In fact the town had less than a century to live.

[2]

PART TWO: FIRST CENTURY POMPEII

The people of Pompeii didn't use our year numbering system. When Romans talked about a year they referred to it by the names of the consuls who had taken office on January 1, so our first century AD began in the year of Gaius Iulius Caesar and Lucius Aemilius Paullus.

Obviously this system wasn't much use for dealing with multiple years so an alternative system existed, counting years from the legendary date of the founding of Rome. The year 1 AD translated to the year 754 AUC (Ab Urbe Condita – From the founding of the city.) Centuries were less important to the Romans and for them 1 AD wasn't the beginning of one anyway, To us, however, 1 AD marks the date Pompeii entered our own era. In fact it was already a surprisingly modern town.

The Wealth of Pompeii

Pompeii's economy had been built on its port, which linked it to the rich trade routes of the increasingly Romanized Mediterranean. Being part of the Empire brought other benefits though. The Romans built their wealth and military power on superior technology, which allowed them to defeat and conquer much larger states in the early years then maintain control of their vast empire. That superiority was applied to every aspect of their lives. Farming in Rome was far more productive and efficient than elsewhere. Low level subsistence farming is inefficient; small patches of land produce only enough to feed the owners and perhaps a small surplus that can be bartered. The Romans operated large farms aimed at producing a single crop for sale and this allowed far greater output. A hundred small plots might feed five hundred people; use the same land for three or four large farms, each producing a single crop, and it could feed three or four thousand. The soil around the Bay of Naples was extremely fertile and Roman farmers were eager to take advantage of it. In fact the fertility was a by-product of Vesuvius's repeated violent eruptions, but the Romans didn't know that. All they knew was that things grew well on the coastal plain around the mountain and that a man could get rich if he owned a farm there. Technology helped; for a culture that could transport clean water 60 miles from the hills inland to the naval base at Misenum, irrigating fields was child's play. Roman farms were barely affected by droughts that could reduce whole nations to famine.

The Romans shipped most of their wheat in from North Africa – the new Roman city of Carthage, built over the rubble of the obliterated enemy, was the center of an enormous grain industry – but large quantities were also grown in Italy. Vegetables couldn't survive the long trip by sea and almost all of them were grown locally. Many familiar vegetables simply didn't exist in ancient Rome; potatoes, peppers and tomatoes – the staple of modern Italian cooking – all come from the Americas and wouldn't reach Europe for another 1,500 years. Others did, but in forms we might not have recognized. The Romans ate carrots but it would take hundreds of years of selective breeding before they were the familiar orange color; the Roman ones came in several colors, but orange wasn't one of them. Leeks were popular and onions formed the base of many meals as well as being used medicinally. Beans were used in soups and stews. Fruit was eaten as a snack or turned into desserts.17 Other farms produced eggs, dairy products and meat. Apiaries scattered between farms supplied honey, which like onions and garlic had medicinal value and was also used where we would use sugar. The land around the Bay of Naples supported a wide range of farms and their produce, along with fish and other seafood from the Bay, was sold both wholesale and retail in Pompeii's vast food market. A large part of the area's agriculture fed local industries; olives and grapes were massive cash crops.

Industry was another area where Rome led the ancient world. Barbarians tended to make their own goods and possessions, and those who had exceptional skills would produce a surplus to trade. More advanced cultures like the Greeks or Celts had cottage industries that could supply the local area with manufactured goods. The Romans on the other hand had large factories that churned out goods in large quantities for distribution and sale throughout the Empire. The Villa Regina at Boscoreale, which was once a suburb of Pompeii, was surrounded by vineyards. It's safe to assume the owners weren't making wine for their own consumption because their storage facility has survived almost intact. A walled courtyard behind the villa contains 18 immense wine jars, buried up to their necks to insulate the contents from the warm climate, with a total capacity of more than 2,600 gallons.18 The remains of the wine press can still be seen, too. Other villas and farms in the area had their own wine presses, or massive olive presses made from volcanic rock. Thousands of gallons of wine and oil were produced and carted into Pompeii for sale, both to meet the town's own huge demands and for sale outside the area. Brands existed and used advertising to boost sales; a popular local wine was Vesuvinum.2 Not everyone was impressed though. The admiral in command of the naval base complained that "Pompeian wines are rather dangerous as they may cause a headache which lasts till noon on the following day."19

Bread was a staple food throughout the ancient world, as it is today, and in most cultures every home had a small domed clay or brick oven to bake loaves. In Rome it was different. People had jobs, and didn't want to start baking when they got home. It was much easier to buy bread from a bakery and Pompeii had over 30 of them. Wheat, both imported and grown locally, was ground into flour in donkey-

---

[2] The name is a combination of *Vesuvius* and *Vinum*, the Latin word for wine.

powered stone mills then batches of loaves were baked in humongous brick ovens.

Romans liked to season their food and the most popular sauce in the Empire – the ketchup of a world without tomatoes – was garum. This was similar to Asian fish sauce; it was made by layering shredded fish with sea salt in a waterproof tank then leaving it in the sun to ferment. Large homes sometimes had their own garum tanks but for anyone without kitchen staff that was too much work, so factories produced it in industrial quantities. Much was used locally, but the sauce made in Pompeii and nearby Herculaneum was extremely high quality and it was widely exported. Garum jars from Pompeii have been found in France.[20] Garum itself was a clear liquid, collected from the top of the tank with a ladle then filtered to remove bones and scales. The remaining mixture, known as allec, was seen as lower quality and was sold cheaply. The diet of poor Romans was based around polenta, with vegetables or meat added when they could afford it, and allec added some flavor to the bland porridge.

The Romans kept a wide range of livestock including millions of sheep. These were valued for their wool, which was the main fiber used for clothing and other products. Unlike the rough fabrics used by barbarians Romans valued high quality cloth, so wool technology was quite advanced. After the sheep were sheared the raw wool was processed by fullers, who washed it in alkaline baths to clean it, wash out the oils and fluffing up the fibers. Human urine contains ammonium salts and made an ideal fulling solution, and also had a bleaching effect which Roman laundries took advantage of. Factories involved in wool processing left tubs outside which acted as public toilets as well as providing a supply of urine, and residents could leave urine buckets outside to be collected. Bulk urine was so pivotal to Roman industry that it was taxed. Faeces were collected too, for use in tanning leather. This sounds fairly disgusting but in fact woolen cloth (and leather) were thoroughly washed after processing to eliminate any smells. Romans valued personal hygiene and would not have tolerated smelling like a sewer. A useful side effect was that the streets of Roman towns weren't awash with the contents of toilet buckets, which most other cultures simply threw out the window as late as the 19th century. This made the Empire's urban areas more pleasant and far healthier.

A Night on the Town

Pompeii offered its residents many of the same amenities as a modern town. As well as the markets that could be found in every culture there were stores that operated much more like a modern one, often specializing in one type of product that might be imported from anywhere in the Empire. The average Roman household had a large number of possessions – clothes, jewelry, cooking utensils, furniture, storage jars, tools and many other manufactured items. A family of nomads in the Middle East, or a hunter in the German forests, could make all the items they needed themselves. This was impossible for the richer and more advanced Romans, so there was a large retail sector.

The Romans also enjoyed active social lives and there were many options for them to choose from. Pompeii had two theaters which put on shows based on Greek drama. The theater was an extremely popular source of entertainment and politicians trying to boost their popularity put on shows much more often than they arranged gladiator fights or chariot races. Cost probably had a lot to do with this but the theater also reached a wider audience; women (and many men) from higher class families tended not to watch the more violent sports, but the theater appealed to almost everyone.

There was plenty of food on offer along Pompeii's streets. Taverns served a variety of meals but for anyone who just wanted food in a hurry there were dozens of fast food restaurants. To anyone familiar with a modern burger franchise the technology will look ancient but the concept will be astonishingly familiar. Tile or marble counters separated the customer area from the kitchen, and a row of lids on the counter concealed the mouths of large ceramic jars – dolia – built in to them.21 These held food or drinks, prepared in advance and ready to be served immediately.

Bars in modern cities tend to cluster around main streets and popular attractions, and Pompeii was no different. Over 200 bars and inns have been identified in the town, with many of them on the streets leading to the amphitheater. Inns offered rooms for guests as well as meals, and the standard of both varied according to price. Meals could range from a bowl of the day's soup, which simmered constantly in a huge pot in the corner, to elaborate multi-course meals. The kitchen of one Pompeii bar contained bones from pigs, sheep, chickens and cattle. Analysis of kitchen waste shows that pork was the most popular meat in ancient Rome and beef was eaten only rarely. Eggs and fish were consumed in large quantities, though. Next to the bar was the House of the Vestals, one of the town's largest and most luxurious mansions. Its occupants ate mostly the same food as was served in the bar, but using higher quality ingredients.

Bars, inns and fast food restaurants all served alcohol; in Rome drinking was socially acceptable and children would often be given diluted wine to drink at a young age. Almost all Roman alcohol came in the form of wine – other drinks like beer were known but not popular – and there was an incredible variety of type and quality. Some wines were served straight, others would be watered or flavored with spices. A popular variation was mulsum, white wine sweetened with honey. Up market inns would serve drinks in fine glassware, with the paper-thin Nile glasses from Egypt being the most stylish. Cheap taverns often used earthenware cups.

A visit to the baths was another popular activity. Pompeii had four public baths and they would have been busy until late at night. Cleanliness was crucial to the Romans and few people had the space or money to have a bathtub at home, but the public ones were cheap – emperors often paid for free access on holidays – and effective. Most baths had a series of pools with cold, warm and hot water, and steam rooms where bathers would scrape dirt and sweat from their skin with a curved tool called a strigil. Some larger ones had private bath rooms which could be hired for parties, and staff could bring in wine, food or even prostitutes. Most baths had at least a small restaurant or tavern as well as other facilities such as barbers, gymnasiums and even public libraries.

The early Empire was a period when people were extremely liberal about sex, but there were limits. Having a relationship with an unmarried female citizen was forbidden – unless she was a prostitute. The ruins of 26 brothels have been found in Pompeii, many of them decorated with erotic paintings and frescoes. There was little or no social stigma attached to visiting one; both prostitutes and clients often carved their names on the walls, sometimes adding improvised verses, and brothels advertised openly.

The number of bars and brothels in Pompeii seems excessive for a town of that size, but there's a simple explanation. The area's wealth and scenery had turned Pompeii into a tourist resort. As well as its farming and industry the town now made a good income from visitors. Many of the Roman aristocracy owned second homes around the Bay. The middle classes could rent a house for the summer or stay in a hotel. Even the lower classes of Roman citizen could afford a room in a cheap inn for a few days, and enjoy a break beside the sea.

The Imperial Machine

Because Pompeii was a Roman town there were plenty visible signs of the Empire's power. In the town center was the Forum, where local politicians and lawyers gave speeches and voted on town laws.

The barbarians outside Rome's borders might have been at the mercy of tyrannical chieftains or scheming priests but here in civilization law and order ruled. The city had police, jails and a neighborhood watch force – the vigiles – which watched for crime or outbreaks of fire. Courts with prosecutors, defenders and judges dealt with anyone accused of a crime, and punishments were laid down by law.

The military was also a constant presence. By the first century AD the Mediterranean, once ruled by the feared Carthaginian enemy, had become what Romans affectionately called mare nostrum – "Our sea." The North African pirates had once preyed on merchant shipping but had finally overstepped the mark and taken the young Julius Caesar hostage.

He had paid them to release him then told them that he would raise an army, return and destroy them.

The pirates laughed and let him go; Caesar raised an army, returned and destroyed them. The sea was safer now and the fleet was reduced in size, but it still had dozens of ships and thousands of sailors and marines. If another enemy ever did threaten Rome from the sea the classis Misensis would be the first line of defense, so it was an important part of the Imperial military.

Meanwhile the men maintained their ships, trained in seamanship and tactics, and spent their money around the Bay.

The townspeople had no fear of them. This was no band of barbarian warriors living off the land; they were paid, received rations and their behavior was controlled by the rigid discipline of the legions.

Across the water from Pompeii the fleet's brilliant but eccentric admiral sat in his office in the large naval base, writing books on science and philosophy.

This, then, was Pompeii in the late first century. It was a town that had grown to around 20,000 people and supported itself with a diverse economy.

The technology belonged to the late Iron Age but apart from the prevalence of slavery the society resembled our own in many ways.

The political system wasn't perfect, but in general law and order was maintained and the economy kept growing.

Women had more rights than almost anywhere else in the ancient world and some of them had a degree of political power.

Racism was rare, although discrimination against non-citizens was rampant.

It was a sophisticated, wealthy town and by the standards of most of human history a very pleasant place to live.

All that was about to end.

THE EARTHQUAKE

After the 1919 season, black baseball was in a state of transition. Team owners realized that they needed more consistency; fans in a city needed to have a home team

they could rely on, with more tangible rivalries and identifiable stars. To remedy this, Rube Foster met with the

The Romans knew about volcanoes and they knew that Vesuvius was one, but they thought it was extinct because it hadn't erupted since records began. In fact the mountain had been quiet for longer than ever before or since – since at least 1,200 BC. The lower slopes were covered with fields and vineyards, and the summit was a flat plateau surrounded by cliffs. In 73 BC a rebel gladiator, Spartacus, led an army of freed slaves to the summit and held out against a Roman militia all summer before defeating his besiegers and escaping. He'd been able to sustain a camp of several thousand men because the summit was well supplied with springs and food was accessible lower down the mountain. There were no signs at all that magma still boiled below the placid peak of Vesuvius. It was earthquakes the residents of the surrounding area feared, not volcanoes.

On February 5, 62 a severe earthquake struck the Bay, causing extensive damage to the coastal towns. Pompeii was among the hardest hit. Earthquakes are common enough in Italy, because it's near a plate boundary. Most of them are caused by the release of tension that's built up between the moving slabs of rock. This one probably wasn't, but that didn't matter to the citizens of Pompeii. What was important was that large parts of the town suffered from the force of the quake. Geologists have analyzed the damage to the town and calculated that the quake had a magnitude of about 6.1, and a series of aftershocks continued for several days. Among the damaged buildings were the Temple of Jupiter and the Vesuvius Gate. Lucius Caecilius Iucundus, the son of a freed slave who had risen to be a successful banker, probably died in the earthquake; his business records, very detailed until then, stop abruptly. His house survived though and someone, probably one of his sons, had a series of bas-relief panels carved to decorate the household shrine. They commemorate the earthquake with images of the damaged temple.

Earthquakes would have been familiar to the people around the Bay of Naples, and no matter how severe this one had been it wasn't going to frighten them away. The citizens of Pompeii began rebuilding the city, although work was slow in some areas – it's believed that further quakes, including one in 64, caused delays. Much of the damage was still visible 17 years later. The town had survived the earthquake but it had been a memorable disaster, and the death and destruction would have come quickly to mind in 79 AD when the ground began to shake again.

The Mountain

The surface of the Earth is made up of dozens of plates – eight of them large – which float on the semi-molten rock of the mantle beneath them. These plates are constantly in motion, driven by convention currents in the mantle. Where neighboring plates are moving apart hot magma forced up from below creates new rock, usually along spectacular underwater mountain ranges like the Mid-Atlantic Ridge. Where they collide, however, enormous destruction can result. The plates each weigh countless trillions of tons, so although they're slow they still have tremendous energy. These enormous slow motion impacts usually force one plate down below the edge of the other in a process called subduction. It's extremely disruptive; huge mountains can be forced up as plates of rock 20 miles thick are buckled and distorted by the strain. The Himalayas were formed by a collision like this, and are still slowly rising as India moves north. At the same time as rock is forced upwards to create mountains the edge of one plate is sinking deep into the hot mantle where it softens and finally breaks up. The descending rock was usually once part of the sea floor and it carries enormous quantities of water down with it. As the rock sinks this water is heated to the point where it becomes a supercritical fluid, an extremely strange state that's neither gas nor liquid.

The rock in the upper layers of the mantle is hot – up to 1,600°F (900°C) – but the enormous pressure at that depth prevents it from melting. It's soft and can flow slowly but it's not a liquid. Saturate it with supercritical water and things change though. Pockets of rock and water have a lower pressure than the surrounding rock, so the melting point drops sharply. Finally it drops below the actual temperature and the rock melts into magma. Because magma is less dense than solid rock it begins to rise, until finally it's trapped by the hard, cool layer of the crust. Sometimes it finds a weak point and forces its way to the surface. That's when a volcano forms.

About 30 million years ago the African plate, moving north at about an inch per year, collided with the Eurasian plate and began to slide beneath it.[22] The boundary between them runs roughly east-west through the southern Mediterranean, with a jagged dogleg up towards the southern coast of Italy. Volcanoes rarely form along the actual boundary of a subduction zone though. Instead they appear along the "magmatic arc" between 50 and about 200 miles away from the boundary, where the hot magma forces its way through the over-riding plate. In this case the result is a region known as the Campanian Volcanic Arc, a loose string of half a dozen volcanoes that crosses the Campania region and runs out into the Mediterranean. It's not a very large arc compared to the infamous Ring of Fire that surrounds the Pacific, and most of the volcanoes in it are either dormant or extinct, but one of them isn't and it makes up for the rest. That's Vesuvius, and most volcanologists think it's the most dangerous volcano on Earth.

There are different types of volcanoes. The ones found near subduction zones are often of the type known as shield volcanoes. These pump out molten lava, often throwing spectacular fountains of glowing liquid rock from their summits. It's a terrifying sight but not actually dangerous. Streams of lava flow down the slopes at a crawling pace, rarely more than a foot a second. They can flow for many miles but eventually they cool and solidify, forming new rock. Shield volcanoes take their name from their shape – they're broad, flattened cones that look a bit like a giant Viking shield. Many islands are shield volcanoes. Hawaii sits over a "hot spot" in the Earth's mantle, a huge plume of hotter magma that endlessly burns away at the crust, and each of its islands is the tip of a giant undersea lava mountain. The Pacific plate is slowly moving and the hot spot isn't, so a trail of older volcanoes, long extinct, straggles away to the northwest. The older they are the more the sea has eroded them, and the further their own immense weight has forced the crust down, so most are now totally submerged. The peaks of a few still protrude above the ocean – Gardner Pinnacles is the last – and others, like Midway, are drowned summits crowned by coral reefs that reach the surface. The surviving islands might look tiny on the surface but actually they're huge mountains rising from the ocean depths; Midway is the peak of an extinct volcano two and a half miles high and over 50 miles across its base. Shield volcanoes are massive. They're not exceedingly dangerous though. They erupt frequently but the hazards are predictable and easily avoided. Occasionally they destroy property but the risk to human life is slow.

Vesuvius isn't a shield volcano. It's a much more dangerous species known as a stratovolcano. Instead of a flattened mound of solid lava these are made up of many layers of different substances – lava, pumice, ash and shattered rock. Their internal structure is complex and irregular; the thick, viscous lava they sometimes belch out flows slowly and cools quickly, building steep, towering cones. Stratovolcanoes are unpredictable and violent. The confused layers within them form traps and blocks that bottle up magma until events far underground build up enough pressure to blow the block apart and open a path to the surface.

Deep below the peak of Vesuvius – more than six miles down into the crust - is a magma chamber, kept full by molten rock forced up from the mantle. As long as the central pipe of the volcano is blocked the magma stays in the chamber. It isn't just sitting there though. Slow chemical reactions are taking place in the searing liquid. As it cools rocks with a higher melting point crystallize out and the concentration of gas dissolved in the magma increases dramatically. The effect is like shaking a bottle of champagne. Nothing much changes on the outside, but sooner or later the cork is going to pop.

Of course that's why champagne corks are securely held with wire. You can't wire down Vesuvius.

As the pressure inside the magma chamber rises the rock above it takes the strain. That has no obvious effect so long as all the activity is miles down in the crust, but if magma starts to escape from the chamber and rise up the pipe to a block closer to the summit signs begin to appear. The immense pressure can be enough to actually distort the entire mountain. The sides of Vesuvius bulge slightly. Huge stresses build up under the slopes. Sometimes the rock cracks or shifts, releasing the stress. The result is a localized – but often extremely violent - earthquake.

It's impossible to say for sure what caused the 62 AD earthquake but Vesuvius has to be a prime suspect. Following the quake 600 sheep were found dead near Pompeii. Dead animals are often an indicator that an apparently dormant volcano is coming to life and sending toxic gas trickling up to the surface. These gases are usually heavier than air and tend to pool in low ground, sometimes trapping and killing livestock. Eruptions in Iceland often leave sheep dead from carbon dioxide asphyxiation.23 In fact there aren't many things that can kill whole flocks of sheep without leaving a mark on them so it seems likely the disturbance of the earth was caused by a buildup of pressure deep below the mountain.

Nero is most famous as the emperor who fiddled while Rome burned, although in fact he was 40 miles away in Antium when that disaster occurred24 and after the fire he offered his palaces as accommodation for people who had been left homeless. He did have some musical talent though, and enjoyed performing in the theater. Because of his tyrannical reputation – the historian Tacitus accused him of having Christians burned in his garden as novelty torches – the unwritten rule was that nobody should leave the theater while the emperor was on stage. This nearly had disastrous consequences right at the beginning of his theatrical career. During his debut performance, in Naples one night in 64 AD, the theater began to shake. Nero sang on. The audience sat, watching nervously as the vibration worsened, until the emperor had finished his song.25 As soon as he was done they headed for the door. Not long afterwards the weakened building collapsed on the empty seats.26 The source of the earthquake, again, was almost certainly Vesuvius.

The geological structure inside Vesuvius is a mess. Ancient lava flows have laid down thick layers of hard, impermeable rock that seal in gas and magma. Mixed through it is volcanic ash – actually pulverized rock and crude glass – and layers of other ejected material like tuff, pumice and tephra. The layers aren't even. The mountain has built up a series of hollow domes over millions of years, which have collapsed one after another, and the giant shattered slabs that remain could almost have been designed to block the central pipe. Just to make it worse cooling lava solidifies into plugs that seal off the last gaps around the broken cone. Each eruption leaves more million-ton debris behind it, choking the throat of the mountain and making the next explosion inevitable. The activity won't stay deep forever. Sooner or later the pipe will fill up and Vesuvius will come to thundering life again.

As magma escapes from the deep chamber and rises up the pipe it changes more rapidly than before. With the pressure of miles of rock gone, gas bubbles out of solution and expands. The pressure rises steeply. Now it's blocked high in the pipe, separated from the surface only by the wreckage of previous eruptions. There are still millions of tons of rock in its way but the huge energy contained in the magma is starting to escape into the mountain. The flanks of Vesuvius swell imperceptibly. Stray trickles of deadly gas find their way out and drift invisibly downhill, perhaps dispersing into the atmosphere, perhaps silently filling a small valley or hollow and turning it into a trap. Springs dry up.27 Streams suddenly change their course. Ten billion tons of stressed rock splits and moves half an inch, slamming the Bay with an energy pulse equivalent to a thousand tons of high explosives. Beginning on August 20, 79 AD the ground shakes repeatedly under Pompeii. The pipe is full now and the vast force of the pressure is trying to bleed off and dissipate - all this activity is caused by the tiny fraction that succeeds. Most of it goes nowhere though; the mountain holds it in. For now.

[3]
Part Three: Death in August

The people of Pompeii certainly hadn't forgotten the 62 AD earthquake; signs of the damage it had caused were still all around them. When a new series of tremors began on August 20, 79 most people tried to ignore them, but as the small quakes continued some grew alarmed enough to think about leaving.

Over the next three days many people took the decision to pack some things and go, at least temporarily. Some archaeologists believe that up to a third of Pompeii's population evacuated the city. At first the departing refugees probably came to no more than a trickle but as magma pushed up the volcano's feeder pipe and the quakes came more frequently their numbers likely swelled until thousands of people were leaving. It was still the fear of another significant earthquake that frightened them though. Nobody had any concerns about Vesuvius.

THE EXPLODING MOUNTAIN

Gaius Caecilius Cilo, 18 years old in August 79, spent most of his time studying rhetoric at a private college in Rome. His tutor was the famous Quintilian, so he was getting the best lessons in logic, persuasion and argument that any student could want, but whenever he got the chance he slipped away to Misenum to visit his favorite uncle. His father had died when he was a child and he'd been brought up by his mother, but shortly after moving to Rome to study he'd begun spending more time with his uncle and the two had come to enjoy each other's company. Gaius Caecilius was a keen student and while he sometimes found rhetoric dull his uncle's work fascinated him. Gaius Plinius Secundus, better known today as Pliny the Elder,3 was one of Rome's leading scholars of science and natural history. He was also politically well connected and a respected military leader, who'd fought in Germany as a junior infantry officer then risen to command armies in France, North Africa and Spain.

In early 79 the emperor Vespasian, who would die after a short illness that June, was looking for a new commander for the fleet at Misenum. Vespasian, like his friend Pliny, came from an Equestrian (lower aristocratic) background and had risen through the ranks of the army. He decided that Pliny would make an ideal admiral –the Romans viewed warships as mobile bits of battlefield for marines to fight on, so their naval commanders were usually also infantry officers – and offered him the job. Pliny accepted and headed off to Misenum, from where he immediately started sending letters describing the scenery, natural history

---

[3] In his will Pliny officially adopted his nephew, who then took the name Gaius Plinius Caecilius Secundus. Today he is known as Pliny the Younger.

and towns. His nephew quickly got into the habit of visiting him whenever he could, and chance had dictated he should be here on August 24, 79 AD.

The naval base was 18 miles from Vesuvius and 20 from Pompeii, so Caecilius couldn't see much of what was happening across the Bay. He spent the morning studying and talking to his mother, who had travelled to Misenum with him; the ongoing earthquakes could be felt faintly but there didn't seem to be anything to concern him much. Then, early in the afternoon, that changed suddenly and dramatically.

The feeder pipe running to the summit of Vesuvius was absolutely full now, and the immense pressure of magma surging up from below and the gas struggling to escape from the molten rock was putting intolerable stress on the mountain. It's likely that the summit was already cracking, throwing out small showers of ash and alarming some of the coastal residents, but much worse was to come. Around 1:00pm the rock plug that had been holding the pressure back failed catastrophically. The magma erupted upwards; as the pressure was released the remaining gas dissolved in it instantly bubbled out of solution and expanded, blowing the top of the volcano apart in a huge plume of shattered rock and foaming lava.

In Misenum Caecilius's mother saw a strange cloud rising inland and asked Pliny - who was working with his books – to look at it. Pliny and his nephew rushed outside to see what was happening and were confronted with a terrifying sight. A grey column towered above the mountain, spreading into a cloud at the top. The young man later wrote in a letter to Tacitus, "I cannot give you a more exact description of its appearance than by comparing to a pine tree; for it shot up to a great height in the form of a tall trunk, which spread out at the top as though into branches"[28] It certainly was reaching a great height; modern volcanologists estimate the plume reached up to 18 miles. When its violent upward momentum finally ran out the lighter material began to spread sideways, mostly downwind – towards Pompeii.

THE FLEET

Pliny, with his interest in natural history, quickly realized that this was a massive volcanic eruption. His curiosity compelled him to investigate what was happening but Misenum was too far from Vesuvius to see much. He had to get closer to the mountain. Luckily he had transport available – after all he commanded the largest fleet in the Roman Empire. Now he issued orders to prepare a fast, light cutter and invited his nephew to come with him. The young man declined; he had work to do, including a writing task for Pliny himself.

At this point Pliny saw the eruption as an interesting natural phenomenon that might make a good chapter in one of his books, but in Pompeii it was rapidly becoming a crisis. Vesuvius was blasting out material at an incredible rate; over 1.5 million tons of searing hot debris was being hurled into the sky every second. The lightest material was rising for miles on a jet of superheated steam and drifting southeast on the prevailing winds, but heavier pieces were falling a lot closer to the mountain. Most of what was erupting from the volcano was pumice, a light rock – it even floats on water – formed when frothing lava cools down in the air. Mixed in with it were large quantities of volcanic ash, fragments of shattered rock. Now chunks of pumice up to an inch wide were falling on Pompeii in a dense rain. Some of the people who had ignored the earthquakes now began fleeing the town, scattering into the countryside to the northeast where the bombardment of stone was less intense. The storm of pumice was dangerous; the chunks weren't heavy but they were falling at some speed, and they were hot. Anyone outside in it would suffer bruises and minor burns at least and also faced the risk of serious injury from collapsing structures or larger chunks of rock thrown out by the volcano. Most of those who fled managed to get away but the majority of the population, sheltering in their homes or workplaces, felt that staying under cover was a safer option. It wouldn't have been a comfortable option though, as the steady buildup of hot pumice pushed the temperature up. By the middle of the afternoon the roof tiles had been heated to a blistering 280°F, hot enough to cause chemical changes that archaeologists could detect many centuries later.29 It would have been hot in the buildings, but not dangerously so. The chance of escape was receding though. As the hours passed the pumice continued to build up, on the roofs and in the streets. By the time one or two feet had blanketed the town it would have been much harder to get out. Because the pumice was light and made up of roughly spherical pellets it didn't form a perfectly solid surface, and anyone trying to cross it would have found their feet sinking in and slowing them down. It would have been worse than walking on sand – more like wading through a children's ball pit, but with rough hot balls that scorched and tore the skin.

Back at Misenum Pliny was still planning a scientific study of the eruption, but before his ship was ready to sail out into the Bay a mounted messenger arrived. Pliny had many friends among the local aristocracy and now one of them was appealing for help. Rectina owned a coastal villa at the foot of the mountain and it was being hammered by the eruption. It was impossible to escape by land, she wrote, so could Pliny arrange a rescue by sea?

Rectina's message is one of the mysteries of that day. Her villa was hours away from Misenum, even on horseback, so if her appeal had been prompted by the explosion it couldn't have reached the admiral before he sailed. It's likely that an earlier, smaller shower of ash had preceded the blast but hadn't been noticed by observers at the base. However it happened the timing was perfect. Realizing that there was a catastrophe looming Pliny swung into action; he roused out the base and ordered that every ship should sail immediately and begin evacuating the settlements along the coast. His own cutter was ready by now and he sailed towards Rectina's house. Rectina had prompted an evacuation that probably saved several thousand people but her own fate, sadly, is unknown. As the ship approached her villa the rain of pumice thickened and heavier, hotter projectiles started splashing down around them. Small earthquakes were creating miniature tsunamis in the Bay, as the sea drew back from the land then crashed ashore again. A new and much deadlier threat was becoming evident, too. From the ship they could see towering clouds that rolled down the slopes of the mountain and obscured the shore. This was the first of the pyroclastic flows.

INCINERATION

Late in the afternoon the column of steam that was pumping the cloud high into the sky faltered temporarily. Instantly the pillar of debris collapsed around the top of the mountain and flowed down the slopes. Tens of millions of tons of pumice, pulverized rock and hot gas was in motion, racing downhill at speeds of up to 450mph. This first surge ran out of energy before reaching Pompeii, but the fishing port of Herculaneum was obliterated and it's likely Rectina's villa was also destroyed. The ships on the Bay were at risk, too. Common sense says that a hot mass hitting the sea will be rapidly extinguished, but the reality is bizarre and horrifying. The heavier rocks splash into the water, instantly vaporizing huge quantities of steam. This supports the remains of the flow, now consisting of smaller fragments and gas, and instead of stopping and cooling down it actually accelerates.30 When Krakatoa exploded in 1883 one flow crossed 30 miles of water and hit the Sumatran coast. Pliny and his fleet were in real danger of being incinerated. Now Pliny's navigating officer was urging him to turn back, but Pliny was determined to achieve something. "Fortune favors the brave," he said, then ordered the ship to head for the home of Gaius Pomponianus.

Pomponianus was a friend of both Tacitus and Pliny, and went on to have a distinguished political and military career. Right now he was in trouble, though. The flow of steam from the underground magma chamber had resumed and Vesuvius was again pumping out a towering cloud. His villa at Stabiae, a few miles south of Pompeii, was cut off by the falling pumice and starting to suffer damage. Pomponianus had his own ship and he'd ordered the crew to get ready to sail, but the strong northwesterly wind that was driving the cloud towards Pompeii had also trapped the ship in his private harbor. Pliny's hope was that his more agile warship would be able to get in, rescue his friend then escape again. Meanwhile the fat admiral sat on deck, observing everything that happened and making endless notes. If he got out of this, he might have been thinking, the next edition of his encyclopedia would have the best description of a volcanic eruption ever written.4

Reaching Pomponianus's villa Pliny found that it wasn't actually in any great danger for the moment, but Pomponianus and the other occupants were close to panic. The admiral decided it was time to set a magnificent example, and he began reassuring the civilians that the situation wasn't as dastardly as they thought. He took a bath – a favorite pastime of his – then ate supper, relaxed and chatty the whole time. Then the group went out onto the terrace

---

[4] Pliny's *Naturalis Historia*, or *Natural History*, was the first encyclopedia and all modern ones follow its style.

to watch the ongoing eruption. It was dark now. The moon was over half full and high in the south, but it was utterly blotted out by the pall of Vesuvius. The night, Pliny wrote, was darker than anywhere on Earth. It shouldn't have been possible to see the volcano, but the towering mountain was periodically illuminated by flames that surged down its slopes. That was nothing to worry about, Pliny consoled his companions; it was just villages – surely abandoned by now – burning. There was no danger.

Pliny might have been mistaken. On the other hand he might have been lying – if anyone in the ancient world understood what a pyroclastic flow was it was probably the intellectual admiral. What's certain is that the flames on Vesuvius weren't farms burning. The eruption was changing its pattern, from the explosive column of the afternoon (what's now known as a Plinian eruption in his honor) to a Pelean phase, where the volcano vomited out heavier, denser clouds in a series of titanic belches. Too heavy to rise on the column of steam, this ejecta spilled over the edge of the crater and raced downhill as a crushing wall of superheated devastation. At least two flows struck Herculaneum during the night, laying down a thick carpet of rock fragments that eventually buried the port under 75 feet of ash and pushed the coastline hundreds of yards further out into the Bay. Between flows the plume thundered into the sky again, resuming the bombardment and pinning the survivors of Pompeii in their homes. The pumice continued to thicken over the doomed town and by the morning of August 25 the streets were choked by over nine feet of it. Escape was now all but impossible and roofs were beginning to collapse under the weight. All the people could do was shut themselves in the sturdiest rooms of their homes and pray for the holocaust to end.

THE ADMIRAL

Pliny's years in the army had taught him to sleep whenever he got the chance, and that's what he now did. Retiring to one of Pomponianus's guest rooms he stretched his ample frame out on the bed, and soon servants in the hall outside could hear contented snores. The situation was worsening though, and renewed quakes began to rock the house. Finally they woke Pliny. Looking out he saw that the courtyard outside his room was now slowly filling up with pumice, and it was clear that the window for escape was narrowing. The rest of the group had been too nervous to sleep and were increasingly agitated. Trying to get back to the warship in the dark was too dangerous to contemplate and staying in the house was equally unthinkable, so Pliny began making preparations for an escape at first light. He sent servants to collect pillows and napkins from around the house. Eccentric as this command seemed, he had a plan and at dawn he put it into effect.

The main hazard would be the falling pumice; even that lightweight rock could cause nasty injuries after falling 90,000 feet, so some protection would be needed. That was where the soft furnishings came in. Pliny ordered each of his companions to take a pillow and tie it to the top of their head with a napkin. That, he assured them, would be enough protection to ward off the pumice and prevent serious injury.

Pliny's plan seemed bizarre, but it worked. Taking flaming torches to ward off the day's unnatural darkness, the group picked their way through the wreckage towards the beach and safety. Their salvation was to be delayed, though. On reaching the shore they found that the waves were still high and violent, and until they subsided there was no way for the beached warship to refloat herself and put to sea. They settled down to wait. Pliny was now exhausted by his exertions and lay down on a piece of sailcloth. Twice he asked for water, and drank it gratefully. Slowly the sea settled. The sailors watched the waves, judging when they'd be able to extend the sweep oars and drive the light hull back out into the Bay. Then the roar of another huge convulsion rolled across the landscape, and instantly everyone turned to look at the mountain. The column of ash had collapsed again.

Another pyroclastic flow, the largest yet, burned its way down Vesuvius. It rolled over the scorched plain that had been Herculaneum and out over the water. Its fringes reached Misenum and terrified Caecilius and his mother. The wave of air pushed in front of the flow itself, hot and stinking of sulfur, struck Pliny's companions and scattered them in panic. The admiral cursed and called for two servants to help him haul his obese body upright but as soon as he reached his feet he collapsed again, lifeless. The survivors later told his nephew that Pliny had been overcome by poisonous gas but as nobody else in the group was affected that seems unlikely. Pliny was seriously overweight, and in the previous hours he'd been under a great deal of stress and exerting himself more vigorously than he had done in years. Almost certainly he fell victim to a massive heart attack; when his body was recovered two days later it showed none of the signs of poisoning but looked "more like a man asleep than dead."31 Rome had lost a talented commander, wise statesman and brilliant scholar. Pompeii, four miles closer to Vesuvius than the beach at Stabiae, faced utter annihilation.

The Death of Pompeii

Three previous pyroclastic flows had exhausted themselves short of Pompeii in the night, leaving the town's surviving residents to huddle in their houses. Now the approaching rumble of the fourth firestorm threw the dying town into a final spasm of panic. Parents grabbed their children and scrambled through the pumice-clogged streets, desperately trying to get to safety that was far beyond their reach. A mother carried her baby through the falling rubble as her two older children struggled along beside her. A banker stood in the vault beneath his home looking sadly at a stack of gold and silver coins. Yesterday his wealth had made him one of the city's leading citizens, and today it was worthless. In one of the town's most popular brothels a woman in her thirties cowered beneath her bed, hoping it would protect her from this next onslaught as it had from the endless rain of stone. Terrified citizens struggled and fought at the half-choked gates. At the Herculaneum gate a lone guard stood stoically under an arch; nobody had ordered him to leave his post, so he would stay. Others, resigned, climbed onto the ramparts and looked towards the mountain, determined to face whatever was coming like Romans. They were the sensible ones. A last-moment struggle for survival was pointless. There was no escape now.

A wall of incandescent volcanic gas slammed into Pompeii at over 300mph, carrying a million tons of pulverized rock and cinders with it. As the fiery wind swirled through the streets it quickly slowed and cooled, but even so it was still close to 400°F as it washed over the town. Because the flow was densely filled with heated rock particles it carried far more energy than clean air at the same temperature would have done. In fact that was a mercy for the remnants of Pompeii. For many years archaeologists believed that the victims had been suffocated by volcanic ash, which would have been a slow and agonizing end. It's now known that death came instantly from the heat of the flow.32 Men, women and children collapsed in the street as the storm caught them, frozen into statues as their muscles charred. The power of the volcanic wind erupted through doors and windows to fill storerooms, basements and anywhere else someone might have taken refuge – even under a bed. Almost everything that projected above the layer of pumice was scythed away.

Pompeii's ordeal had lasted over 17 hours from the first showers of pumice. Thousands had fled. Thousands more had died. Now, in a matter of seconds, Vesuvius delivered the coup de grace. People, animals, plants – the surge snuffed out all life as it roared through the city.

The huge volume of rock carried along by the flow settled over the ruins, entombing those it had killed. Two further surges swept across Pompeii later in the morning, but there was nothing left there to kill. More ash blanketed the wreckage, covering the remains of up to 16,000 people. By the time the last clouds dispersed on August 26 and the search for survivors began Pompeii had vanished without trace.

[4]
Discovery and Conservation

After the disaster, the Romans rebuilt around the Bay of Naples, and as centuries passed the new settlements gradually evolved into the modern Italian towns that stand there today. Pompeii was forgotten, its name known only by a few scholars of ancient history. The Roman Empire expanded to dominate all of Europe and most of North Africa and the Middle East, then stagnated, weakened and split. Christianity, a despised cult barely tolerated by the Caesars, became the official religion of the crumbling Roman state. Barbarians looted Rome and the center of the Empire shifted east to Byzantium. History moved on. Beneath the plain of Naples, however, a Roman city lay frozen in time waiting to be discovered.

In 1599, a team of workmen were excavating an underground channel to divert the Sarno River when, deep under a layer of soft volcanic ash, they encountered an ancient wall decorated with Roman frescoes. One of the period's leading architects, the Swiss-born Domenico Fontana, was called in to examine the ruins, and he quickly realized they were Roman. It's likely he also saw an inscription which included the name Pompeii, but Fontana decided not to announce the rediscovery of a lost Roman city. Instead, he covered the walls again and advised the workmen to divert their channel. This was probably a wise move; in the late 16th century, the church was struggling to suppress a rising tide of "heretical" thought and scientific discovery, and their reaction to the erotic paintings and frescoes of Pompeii is easy to predict. Fontana may have intended to hide the images because he found them offensive, but the effect was to prevent an act of cultural vandalism.

The next discovery of Roman ruins came in 1738, but this time it was Herculaneum that was found. Workmen digging the foundations for a new palace stumbled on the ruined fishing port, and this time the find was announced. The Renaissance was in full swing and rediscovered Roman culture was now seen as something to be celebrated, not obliterated out of prudishness, so the discovery caused a lot of excitement. It also had political value, proving that Naples had been an important area so far back in the past. With the renewed interest in the area's ruins, a new search for Pompeii began, and the city was rediscovered once more in 1748. Now serious excavations began, aimed at bringing the ancient buildings back into the daylight.

Most archaeological excavations are relatively shallow, usually involving just a few feet of soil. Pompeii was different. The ruins were covered by up to 70 feet of volcanic debris, so massive quantities of rock and ash had to be cleared away. Over the space of a century, a succession of engineers managed to clear parts of the city. Unfortunately, their techniques were often crude, and it's likely that much valuable information was lost, but they were learning. One thing that had puzzled archaeologists were strange hollows in the ash, each of them containing a single human skeleton. In 1863 Giuseppe Fiorelli had an idea. Next time one of his workmen found one of the peculiar cavities Fiorelli injected plaster into it and let it set, then carefully cleared away the surrounding ash. The hollows were the spaces left behind after the corpses of the victims had decayed, and Fiorelli had produced a lifelike cast showing how the body had looked as the ash buried it. This technique is still in use,33 although resin is replacing the plaster.

Pompeii's ruins quickly began attracting visitors, and by the early 19th century was part of the "Grand Tour" of Europe popular with the upper classes. Now more than 2.6 million people a year visit the ruins. That's terrific news for the local economy because tourism brings in a lot of money and supports a large percentage of the workforce. It's good news for the tourists, too. The sophistication of ancient Rome is hard to appreciate from books, but it's remarkably obvious in Pompeii. In fact about the only thing, the level of tourism isn't good news for is the ruins themselves. The constant traffic through the old streets is slowly damaging them, and the local authorities are trying to persuade visitors to tour Herculaneum and other sites to relieve the pressure on Pompeii itself.

The problem is that the ruins were largely protected for 1,700 years by the ash and pumice that covered them, but as soon as a section is excavated it becomes vulnerable. Rain and wind cause erosion and can damage paintings; tourists disturb the cobblestones and, incredibly, sometimes steal or vandalize items. There's a real danger that unless properly cared for the last traces of Pompeii will be destroyed after having survived everything Vesuvius could throw at them; in 2010 the Schola Armatorum, the House of the Gladiators, collapsed.34 It's believed that heavy rain had damaged the foundations, and some archaeologists have accused local authorities of neglecting the site.

The size of Pompeii and the range of threats to the ruins make conservation a real challenge. As well as tourism and erosion many other factors can cause damage; sunlight fades the ancient paintings, for example. Plants colonize the old town and their roots crumble bricks and tiles. Even early attempts at restoring some buildings to their original appearance have caused irreparable damage. Limited funding restricts what can be done and there's pressure to stop excavating new parts of the city until what's already exposed has been adequately protected. A lot of progress has been made in recent years though, and there is hope for the future.

Conclusion

The culture of ancient Rome has fascinated historians and the public almost since the Empire collapsed. Much of what people knew or believed about it was suspect, though. Ideas were based on written works which had often been copied and re-copied through the centuries, by copyists who may have altered the texts to suit their own agendas. Archaeological finds tended to be scattered objects or lone buildings. Nobody actually knew what a Roman town had looked like because most of them had remained constantly occupied for centuries and gradually evolved into modern towns. Roman buildings had been modified, extended, demolished and replaced; their roads had vanished beneath railways, freeways or fields. Surviving structures like the Coliseum were awe-inspiring, but a long way from what the typical Roman citizen lived in. The literature, heroes and military conquests of the Roman Empire continued to be studied, but the people themselves were almost forgotten.

The rediscovery of Pompeii changed all that. The disaster of August 24-25, 79 AD swept away a thriving, prosperous town and killed thousands, but at the same time it created a time capsule of buildings, artifacts and other objects that together give a detailed picture of life in a Roman town at the height of the Empire's power. While tourists walk the streets and marvel at the rich culture on display archaeologists and anthropologists study household objects, architectural details and even the contents of rubbish heaps and toilets, slowly building a better understanding of the people who lived in and around Pompeii. In recent years, it's become ever more obvious that in many ways those people were a lot more like us than we could ever have imagined.

There's another lesson in the ruins of Pompeii, of course, a darker one that people don't like to think of. The Roman Empire that Pompeii belonged to was the unchallenged superpower of its day, even more dominant than the USA is 20 centuries later. Military and economic control of the entire Mediterranean coastline, and all of Europe as far north as the Scottish border, were concentrated firmly in one city on the Tiber. When a force of nature like a volcano broke loose, however, the Romans' power was helpless.

Perhaps 100,000 people lived around the Bay of Naples in August 79. Around 16,000 of them died. Today there are nearly 4 million people packed into the same area, and they're all living, figuratively and literally, in the shadow of Vesuvius. The brooding mountain has been silent for nearly 70 years, but it's still active. Geologists have better tools than Pliny could ever have dreamed of, including and hope they'll be able to predict a future eruption in time to evacuate the region. Evacuation plans are in place, and it's estimated that everyone could be moved out of the area in seven days. In 79, there were only four days' warning of the explosion but modern tools, including GPS sensors to detect tiny expansions in the mountain and chemical analysis of gases escaping from below ground, may give more notice. We can only hope so because the question isn't if Vesuvius will erupt again. It's when.

NOTES

[1] Spiegel Online, Jul 3, 2008, *Naples Trash Trauma*
http://www.spiegel.de/international/europe/naples-trash-trauma-psychologists-to-counsel-italians-on-garbage-crisis-a-563704.html

[2] The Independent, Mar 19, 2005, *Soviet navy 'left 20 nuclear warheads in Bay of Naples'*

http://www.independent.co.uk/news/world/europe/soviet-navy-left-20-nuclear-warheads-in-bay-of-naples-6150280.html

[3] Strabo, *Geographica*, Book 5, Ch 4 http://www.perseus.tufts.edu/hopper/text?doc=Perseus:text:1999.01.0239:book=5:chapter=4&highlight=pompeii

[4] Livius, Titus, *Ab Urbe Condita*, Book 1

[5] Livius, Titus, *Ab Urbe Condit*, Book 9

[6] Plutarch, *Pyrrhus*, Book 21 Chapter 9.

[7] Plinius, Gaius Secundus, *Naturalis Historia*, Book 15 Chapter 23

[8] Jones, Benjamin, *A CHANGING DEFENSE: Roman Impetus for the Evolution of Pompeian Fortification*, p14

http://www.academia.edu/1231059/A_Changing_Defense_Roman_Impetus_for_Pompeiian_Fortification

[9] Archaeology's Interactive Dig, Aug 7, 2001
http://interactive.archaeology.org/pompeii/field/5.html

[10] Jones, Benjamin, *A CHANGING DEFENSE: Roman Impetus for the Evolution of Pompeian Fortification*, p10

http://www.academia.edu/1231059/A_Changing_Defense_Roman_Impetus_for_Pompeiian_Fortification

[11] Cicero, Marcus Tullius, *Pro Sulla*
http://www.egs.edu/library/cicero/articles/pro-sulla-oratio/ (In Latin)

[12] University of Chicago, *The Amphitheater at Pompeii*

http://penelope.uchicago.edu/~grout/encyclopaedia_romana/gladiators/pompeii.html

[13] Tacitus, Publius Cornelius, *Annales*, Book 14 Chapter 17

[14] US Geological Survey, *Estimated Use of Water in the United States in 2005*, p. 20
http://pubs.usgs.gov/circ/1344/pdf/c1344.pdf

[15] The Conference Board of Canada, *How Canada Performs – Water Withdrawals*

http://www.conferenceboard.ca/hcp/details/environment/water-consumption.aspx

[16] Romanaqueducts.com
http://www.romanaqueducts.info/aquasite/serino/

[17] Apicius, *De re coquinaria*

[18] Claire Benn, *Pompeii and Herculaneum. Economy – Industries and Occupations*
http://history-sjcdubbo.wikispaces.com/file/view/Pompeii%20and%20Herculaneum%20Economy.ppt#256,1,Pompeii and Herculaneum Economy: Industries and Occupations

[19] Plinius, Gaius Secundus, *Naturalis Historia*, Book 14 Chapter 70

[20] Claire Benn, *Pompeii and Herculaneum. Economy – Industries and Occupations*
http://history-

[21] Claire Benn, *Pompeii and Herculaneum. Economy – Industries and Occupations*
http://history-

[22] National Geographic, Apr 19, 2011, *Europe Starting to Dive Under Africa?*

http://news.nationalgeographic.com/news/2011/04/110419-europe-africa-mediterranean-earthquake-risk-increasing-earth-science/

[23] Seed Daily, May 25, 2011, *Near Iceland volcano, farmers rescue animals from ash*

http://www.seeddaily.com/reports/Near_Iceland_volcano_farmers_rescue_animals_from_ash_999.html

[24] Tacitus, Publius Cornelius, *Annales*, Book 15, Chapter 39

http://penelope.uchicago.edu/Thayer/E/Roman/Texts/Tacitus/Annals/15B*.html

[25] Suetonius, Gaius Tranquillus, *De Vita Caesarum*, Nero, Chapter 20

http://penelope.uchicago.edu/Thayer/E/Roman/Texts/Suetonius/12Caesars/Nero*.html

[26] Tacitus, Publius Cornelius, *Annales*, Book 15, Chapter 34

http://penelope.uchicago.edu/Thayer/E/Roman/Texts/Tacitus/Annals/15B*.html

[27] Current Archaeology, Sep 28, 2008, *Visiting Pompeii*
http://www.archaeology.co.uk/cwa/world-features/visiting-pompeii.htm

[28] Plinius, Gaius Caecilius Secundus, *LXV. To Tacitus*
http://www.bartleby.com/9/4/1065.html

[29] Zanella et al, *Influences of urban fabric on pyroclastic density currents at Pompeii (Italy)*

http://www.earth-prints.org/bitstream/2122/2370/1/1245.pdf

[30] Freundt, Armin, *Entrance of hot pyroclastic flows into the sea: experimental observations*

http://cat.inist.fr/?aModele=afficheN&cpsidt=14575991

[31] Plinius, Gaius Caecilius Secundus, *LXV. To Tacitus* http://www.bartleby.com/9/4/1065.html

[32] Mastrolorenzo et al, 2010, *Lethal Thermal Impact at Periphery of Pyroclastic Surges: Evidences at Pompeii*

[33] BBC News, Apr 5, 20120, *Pompeii's frozen victims on display*
http://news.bbc.co.uk/2/hi/europe/8599122.stm

[34] Sky News, Nov 6, 2010, *Pompeii Gladiator Training Centre Collapses*
http://news.sky.com/story/818070/pompeii-gladiator-training-centre-collapses

CPSIA information can be obtained
at www.ICGtesting.com
Printed in the USA
LVHW100232111122
732914LV00005B/449